ADDITION

FACTS

COLOURING BOOK

1-12

THE EASY WAY
TO LEARN THE
ADDITION TABLES

Addition Facts Colouring Book 1-12: The Easy Way to Learn the Addition Tables

ISBN 978-1-897384-82-4

Magdalene Press, 2016

ADDITION FACTS

COLOURING BOOK

THE EASY WAY TO LEARN THE ADDITION TABLES

1-12

1 +

1 + 1 = 2

1 + 2 = 3

1 + 3 = 4

$$1 + 4 = 5$$

$$1 + 5 = 6$$

$$1 + 6 = 7$$

$$1 + 7 = 8$$

1 + 8 = 9

1 + 9 = 10

1 + 10 = 11

1 + 11 = 12

1+12=13

2+

2+1=3

2+2=4

2+3=5

$$2+4=6$$

$$2+5=7$$

$$2+6=8$$

$$2+7=9$$

$$2 + 8 = 10$$

$$2 + 9 = 11$$

$$2 + 10 = 12$$

$$2 + 11 = 13$$

2+12=14

3 +

3 + 1 = 4

3 + 2 = 5

3 + 3 = 6

$$3 + 4 = 7$$

$$3 + 5 = 8$$

$$3 + 6 = 9$$

$$3 + 7 = 10$$

$$3 + 8 = 11$$

$$3 + 9 = 12$$

$$3 + 10 = 13$$

$$3 + 11 = 14$$

3+12=15

4 +

4 + 1 = 5

4 + 2 = 6

4 + 3 = 7

$$4 + 4 = 8$$

$$4 + 5 = 9$$

$$4 + 6 = 10$$

$$4 + 7 = 11$$

4 + 8 = 12

4 + 9 = 13

4 + 10 = 14

4 + 11 = 15

4+12=16

5 +

5 + 1 = 6

5 + 2 = 7

5 + 3 = 8

5 + 4 = 9

5 + 5 = 10

5 + 6 = 11

5 + 7 = 12

$$5+8=13$$

$$5+9=14$$

$$5+10=15$$

$$5+11=16$$

5+12=17

6+

6+1=7

6+2=8

6+3=9

$$6 + 4 = 10$$

$$6 + 5 = 11$$

$$6 + 6 = 12$$

$$6 + 7 = 13$$

$$6+8=14$$

$$6+9=15$$

$$6+10=16$$

$$6+11=17$$

6+12=18

7 +

$$7 + 1 = 8$$

$$7 + 2 = 9$$

$$7 + 3 = 10$$

$$7 + 4 = 11$$

$$7 + 5 = 12$$

$$7 + 6 = 13$$

$$7 + 7 = 14$$

$$7 + 8 = 15$$

$$7 + 9 = 16$$

$$7 + 10 = 17$$

$$7 + 11 = 18$$

7+12=19

8 +

8 + 1 = 9

8 + 2 = 10

8 + 3 = 11

$$8 + 4 = 12$$

$$8 + 5 = 13$$

$$8 + 6 = 14$$

$$8 + 7 = 15$$

8 + 8 = 16

8 + 9 = 17

8 + 10 - 18

8 + 11 = 19

8+12=20

9 +

$$9 + 1 = 10$$

$$9 + 2 = 11$$

$$9 + 3 = 12$$

$$9 + 4 = 13$$

$$9 + 5 = 14$$

$$9 + 6 = 15$$

$$9 + 7 = 16$$

$$9 + 8 = 17$$

$$9 + 9 = 18$$

$$9 + 10 = 19$$

$$9 + 11 = 20$$

9+12=21

10+

$$10 + 1 = 11$$

$$10 + 2 = 12$$

$$10 + 3 = 13$$

$$10 + 4 = 14$$

$$10 + 5 = 15$$

$$10 + 6 = 16$$

$$10 + 7 = 17$$

$$10 + 8 = 18$$

$$10 + 9 = 19$$

$$10 + 10 = 20$$

$$10 + 11 = 21$$

10+12=22

11+

11 + 1 = 12

11 + 2 = 13

11 + 3 = 14

$$11 + 4 = 15$$

$$11 + 5 = 16$$

$$11 + 6 = 17$$

$$11 + 7 = 18$$

$$11 + 8 = 19$$

$$11 + 9 = 20$$

$$11 + 10 = 21$$

$$11 + 11 = 22$$

11+12=23

12 +

12 + 1 = 13

12 + 2 = 14

12 + 3 = 15

$$12 + 4 = 16$$

$$12 + 5 = 17$$

$$12 + 6 = 18$$

$$12 + 7 = 19$$

12 + 8 = 20

12 + 9 = 21

12 + 10 = 22

12 + 11 = 23

12+12=24

Congratulations!

You did it!

www.ingramcontent.com/pod-product-compliance
Lightning Source LLC
Chambersburg PA
CBHW080722220326

41520CB00056B/7375